Differential Drive Robots

A Primer for College Students

Published by Crossbridge Books
Worcester
www.crossbridgeeducational.com

ISBN 978-1-913946-89-0

British Library Cataloguing Publication Data

A catalogue record for this book is available from the British Library

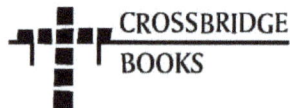

CROSSBRIDGE
BOOKS

Differential Drive Robots

A Primer for College Students

Dr. Colin B. Price

Contents

Page

Preface

What is this book about and who should buy it? This book supports a 2nd year module at the University of Worcester, UK 'Robotics. The module is very hands-on using real robots, real sensors, and experiencing real problems. This short edition is intended for future Worcester students, though I hope that some material may be of wider interest. It concentrates on the theory needed to code working applications on a number of robotic platforms but includes code snippets.

Additional Materials. Teaching resources including worksheets and computer code are available from the author at c.price@worc.ac.uk who will send you a link to the module website. This includes files for laser-cutting a 2-wheeled stepper motor robot.

Software and Hardware used. A number of free pieces of software are mentioned in this book. We make extensive use of Octave in plotting and general data analysis [1], In addition to our in-house robot, we use the Parallax Activity bot [2], supplemented with the Pixy2 camera [3].

Acknowledgements. I would like to thank my past students on this module (2021-22) for their contributions to this book. Their work in class has inspired (and challenged me), and their critical eagle-eye review of my text, as it emerged, was incredibly useful. Thanks to all of you; you know who you are! I would like to thank one of my other students **Flynn Osborne** who worked closely with me in various aspects of design, development and testing the materials used in this book. This included writing computer code, conducting and analysing simulations and constructing physical robots.

References

[1] John W. Eaton, David Bateman, Søren Hauberg, Rik Wehbring (2019). GNU Octave version 5.2.0 manual: a high-level interactive language for numerical computations.

https://www.gnu.org/software/octave/doc/v5.2.0/

[2] https://www.parallax.com/education/robotics/

[3] https://pixycam.com/pixy2/

Chapter 1
Robot Kinematics

1.1 A brief Introduction

Kinematics is the study of movement (well almost) so in this chapter we are looking at the principles of robot movement, for our 2-wheeled differential drive robot. The explanation is mathematical; this is necessary for us to write code to get the robot moving. The most important expressions are highlighted. So how does a wheeled robot move? Simply by driving its wheels. If you drive both wheels at the same angular speed, then the robot moves forward. If you drive the left wheel faster, then the robot will arc to the right. Driving both wheels at the same speed, but in opposite directions, will make the robot spin about its axis.

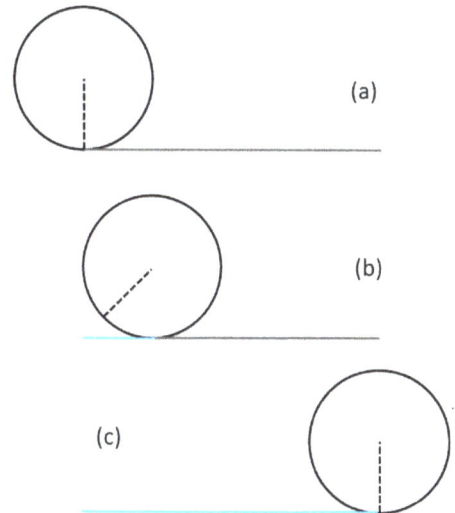

Figure 1.1. Rotating wheel leaving a track of blue paint on the surface

1.2 Linear and Angular Velocities

Let's see how wheels work. Fig. 1.1 shows a wheel completing a full revolution. Imagine the type is coated with paint, so as the wheel rotates it paints a nice line on the surface. How long is this line? Simply the circumference of the wheel.

Now let's do a quick calculation. I guess you will remember the expression for the circumference of a circle radius r. This is of course $2\pi r$. So, if we have a wheel of radius 33mm then one rotation will shift the robot a distance $(2)(3.1415)(33) = 207$mm.

And now for some maths. If one rotation gives a distance of $2\pi r$ then half a rotation will give half of this distance, but what about if the angle of rotation is $\Delta\theta$? (Here the symbol Δ means a change and θ is the angle. Well, the arc of the circle corresponding to the angle $\Delta\theta$ is just

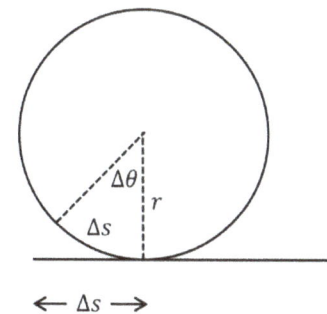

Figure 1.2. Wheel rotating with a change in angle

$$\Delta s = r\Delta\theta \qquad (1)$$

where the angle is in radians (more on that later). So, this is the length of the line painted by the robot, and is the distance moved forward, look at Fig.1.2.

Now let's think about the robot's speed in mm/sec. Speed is the distance moved Δs in a time interval Δt. Here's the expression for speed.

$$v = \frac{\Delta s}{\Delta t} \qquad (2)$$

Continuing with our example, if the robot wheels make one revolution in 2 seconds, then the robot speed is

$$v = \frac{2\pi r}{2} = \frac{207}{2} = 103.5 \; mm/sec$$

Try to imagine what that means. The symbol for speed is v since what we are really talking about is *velocity*, speed in a particular direction (forward or backward).

But robots move by turning their wheel with their servomotors and our computer programs must provide *drive* to the motors to make them rotate. To make a servomotor rotate, we must give it a series of 'pulses' where the pulse width determines the rotational speed. For the robot we shall be using, the 'Parallax Activity Bot /BoE Bot' the relationship between pulse width and speed is shown in the diagram below. The points show real measurements, and the

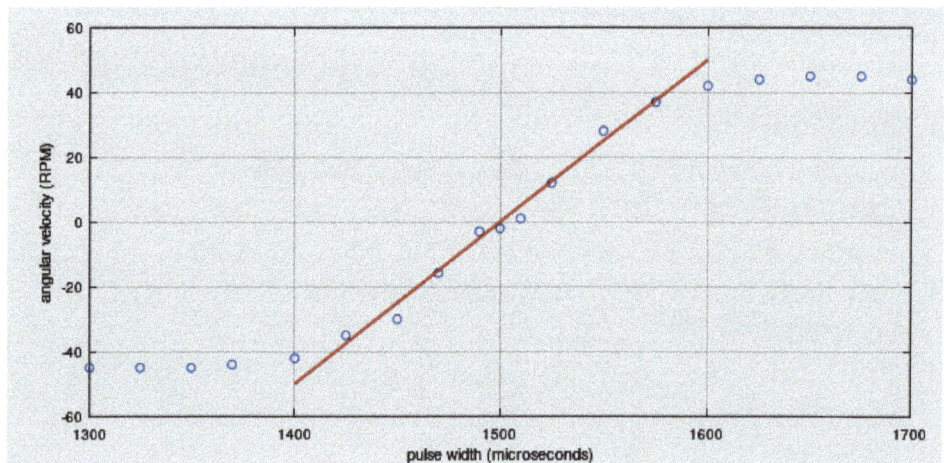

red line has been placed to capture the most useful part of the experimental curve, where speed is proportional to pulse width. You can see that with a pulse width of 1500 µs (microseconds) the motor does not rotate. Therefore, in our code (and in our thinking) we shall define a quantity **drive** *relative to this 1500*. So, a drive of 50 will create a rotational velocity of around 30 rpm, and a drive of -50 will rotate the motor at this rpm, but in the opposite direction.

It's easy to get a relationship between drive and rpm from the red line above; here we find this is

$$drive = \frac{50.0}{30.0} rpm \qquad (3)$$

In our code, you will see the variables **driveL** and **driveR** which are, of course, the drives sent to the left and right servomotors.

Now we need to understand how the values of driveL and driveR will determine the robot's speed; let's assume these are the same, so the robot will move forwards (we'll look at other possibilities later). If we stick expression (1) for the distance gone when the wheel rotates through an angle, into expression (2) which is the definition for linear robot velocity moving forward, we get

$$v = \frac{\Delta s}{\Delta t} = \frac{r \Delta \theta}{\Delta t} = r \frac{\Delta \theta}{\Delta t}$$

Here $\Delta\theta/\Delta t$ tells us how fast the wheel angle changes with time, so this is the *angular velocity* of the wheel; we give this the symbol ω so we have a fundamental expression

$$v = \omega r \qquad (4)$$

This makes sense; if the wheel radius r is increased, then the linear speed v is increased, and if the angular velocity ω is increased (the wheel rotates faster) then the linear speed is increased. All seems good, and it is. There may be a conceptual stumbling block, however, it's due to the *units* of angular velocity ω. The angle change in (1) is measured in *radians* (in one revolution, there are 360 degrees which is 2π radians, a little over 6). So, we need to connect angular velocity in radians/sec to angular velocity in rpm, so we can use the above drive graph.

> **Help! What are radians?**
>
> Think of a wheel rotating once, through 360 degrees. We know the distance gone is the circumference of the wheel $2\pi r$. Now expression (1) tells us that this distance is $r\Delta\theta$. So, we have
> $$2\pi r = r\Delta\theta$$
> and cancelling the r gives us
> $$\Delta\theta = 2\pi$$
> therefore, in a circle, there are 2π radians.

Let's say the wheels are rotating at n rpm, i.e., n_{rpm}. Therefore, the revs per second is

$$n_{rps} = \frac{n_{rpm}}{60}$$

Each rotation the wheel rotates 2π radians, therefore the radians per second (ω) is just

$$\omega = 2\pi\frac{n_{rpm}}{60} = \frac{2\pi}{60}n_{rpm}$$

i.e.,

$$\omega = \frac{2\pi}{60}n_{rpm} \qquad (5)$$

Let's work through an example how we would use this maths to make a robot move forwards a desired distance, in a desired time.

Let's take the case of driving the robot a desired distance of 80mm in a desired time of 2 seconds, a speed of 40 mm/s.

$v = \frac{\Delta s}{\Delta t}$ (2)	$v = 80/2 = 40$ mm/sec	

$\omega = \frac{v}{r}$ from (4) $\omega = 40/33.0 = 1.21$ rad/sec

$n_{rpm} = \frac{60}{2\pi}\omega$ from (5) $n_{rpm} = (60*1.21)/6.28 = 12$ rpm

drive $= (50.0/30.0)*12 = 20$

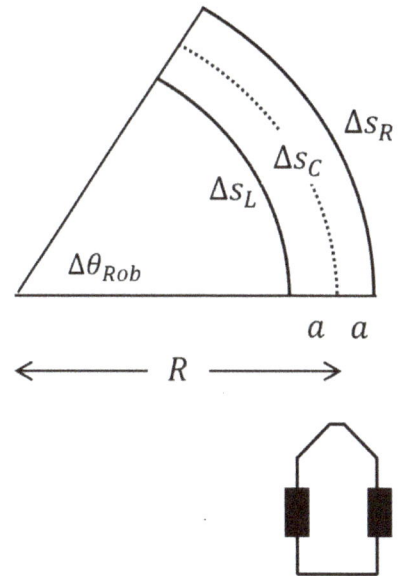

We shall revisit this below when we discuss how to code our robot.

1.3 Movement on an arc

Have a look at Fig.1.3, at the bottom you will see the robot. The length of its axle, connecting its wheels is $2a$ and for the Parallax robot we are using, this is 104.0 mm, so we have $a = 52$ mm.

Figure 1.3. Robot moving on an arc. Distance a is between each wheel and the robot centre.

The diagram shows that the centre of the robot (between the wheels) moves along an arc of radius R. So the left wheel moves along an arc of radius $(R - a)$ and the right wheel moves along a larger arc of radius $(R + a)$. Clearly the right wheel is moving faster than the left. The robot's *pose* changes, it started off moving North, and it ends up moving Northwest, having changed its bearing by an angle θ_{Rob}. The subscript *Rob* reminds us we are thinking about the entire robot, rather than its wheels.

To understand the maths which follows, we apply the relationship
$\Delta s = r\Delta\theta$ used above. So, for the left wheel we have

$$\Delta s_L = (R - a)\Delta\theta_{Rob}   (6a)$$

and for the right wheel

$$\Delta s_R = (R + a)\Delta\theta_{Rob} \qquad (6b)$$

Now let's imagine that the robot takes a certain time interval Δt to complete its trajectory along the arc. To find the robot wheel speeds, we must divide distance gone by each wheel by this time interval, see expression (2). So, we get for the left wheel

$$v_L = \frac{\Delta s_L}{\Delta t} = (R - a)\frac{\Delta\theta_{Rob}}{\Delta t} \qquad (7a)$$

and for the right wheel

$$v_R = \frac{\Delta s_R}{\Delta t} = (R + a)\frac{\Delta\theta_{Rob}}{\Delta t} \qquad (7b)$$

This is fine, but these expressions are not really telling us much. So, we must go forwards a bit. Remember the expression (1) connecting wheel distance and angle. For the left and right wheels these become, (since both wheels have the same radius).

$$\Delta s_L = r\Delta\theta_L, \quad and \quad \Delta s_R = r\Delta\theta_R$$

and putting these into expressions (7a) and (7b) we find

$$r\left(\frac{\Delta\theta_L}{\Delta t}\right) = (R - a)\left(\frac{\Delta\theta_{Rob}}{\Delta t}\right) \qquad (8a)$$

and

$$r\left(\frac{\Delta\theta_R}{\Delta t}\right) = (R + a)\left(\frac{\Delta\theta_{Rob}}{\Delta t}\right) \qquad (8b)$$

I've stuck in some brackets here, where changes in angles are divided by a corresponding change in time. These are angular velocities, speeds of rotation.

The symbol for angular velocity is **omega** ω (lower case) or Ω (upper case). Lower case ω is the angular velocity of the *wheels* and upper case Ω is the angular velocity of the robot body, when viewed from above; this is just the rotation speed of the robot.

So, we can rewrite equations (8) like this

$$r\omega_L = (R - a)\Omega_{Rob}$$

$$r\omega_R = (R + a)\Omega_{Rob}$$

therefore

$$\omega_L = \frac{(R - a)\Omega_{Rob}}{r} \qquad (9a)$$

$$\omega_R = \frac{(R + a)\Omega_{Rob}}{r} \qquad (9b)$$

These are very useful expressions. As with all expressions, think of the stuff on the right of the = sign as an *input* to a computation, what we want to calculate, and the stuff on the left is what we have to code to make this happen. So, if we want the robot to go around an arc of radius R with an angular velocity Ω_{Rob} then we have to make the motors rotate with angular velocities ω_L and ω_R.

1.4 Special Cases

There are two special cases of the maths in expressions (9). First, when the robot is moving straight, then we can write $R \to \infty$ so we can neglect a in the expressions, and R divides out. You can see this by calculating the ratio of the wheel omega's

$$\frac{\omega_L}{\omega_R} = \frac{(R - a)}{(R + a)} \qquad (10)$$

where the ratio becomes $R/R = 1$ which tells us that the omegas are the same, i.e., both wheels rotate at the same speed.

The other special case is when $R = 0$, this is where the robot rotates about its centre. Putting this into expressions (10) we find

$$\frac{\omega_L}{\omega_R} = -1 \qquad (11)$$

so, the omegas are equal and opposite!

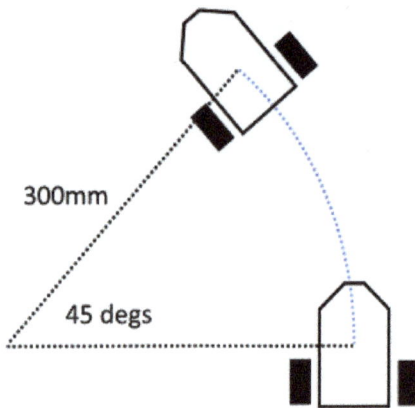

Figure 1.4 Worked example. The robot rotates 45 degs on an arc 300mm and takes 2 seconds.

A Worked Example

Let's say we want our robot to travel along an arc of radius 300mm and change its pose by 45 degrees, and it does this in 2 seconds, see Fig. 1.4.

The angle expressed in radians is $45\pi/180 = 0.785$rad. The angular velocity of the robot Ω_{Rob} is $0.785/2.0 = 0.393$ rad/sec.

Since $a = 52$mm for the Parallax robot and our arc has a radius of 300 mm, plugging these into expressions (8) gives us

$$\omega_L = \frac{(300-52)0.393}{33}, \quad \omega_R = \frac{(300+52)0.393}{33}$$

and so, we calculate

$$\omega_L = 2.85 \; rad/s, \quad \omega_R = 4.19 \; rad/s$$

Now we need to convert these *omegas* to *rpms*. Since an omega of 2π rad/s corresponds to 1 rev/sec, then 1 rad/s corresponds to $1/2\pi$ rev/sec. Then ω rad/s corresponds to $\omega/2\pi$ rev/sec, and therefore to $60\omega/2\pi$ rpm. To get revs per second we divide the omegas by 2π which gives us

left 0.45 revs/sec   right 0.67 revs/sec

and to get the revs/min we multiply by 60

left 27 revs/min   right   40.2 revs/min

and using the expression (3) for drive, we finally have

left drive 45   right drive 67

which are the drive signals we send to our motors.

All of these calculations are done in our Arduino code. The purpose of this worked example is simply to explain what the code actually does.

1.5 How to Code a Real Robot

We need to get the robot to move forward a certain distance we specify or rotate an angle we specify or rotate on an arc. In our code, we have to specify **drives** which will make the robot move as we want, then send these drives to the servos. Here's a code snippet which gets the robot moving forwards

```
driveL = 30;
driveR = 30;
driveServos(driveL,driveR);
```

This code will get the robot moving (for ever) but we would rather like to tell the robot *how far to move* and *with what speed*. So, here's some code to get the robot moving forward for a specified distance (in mm) over a specified time. We relate the code to the corresponding maths. All the variables have been declared for you in the code templates.

`desDist = 80;`	
`desTime = 2.0;`	
`desSpeed = desDist/desTime;`	expression (2)
`omega = desSpeed/wheelRad;`	from expression (4)
`rpm = (60/(2*PI))*omega;`	from expression (5)
`driveL = (50.0/30.0)*rpm;`	expression (3)
`driveR = driveL;`	
`driveServos(driveL,driveR);`	
`delayTime = (int)(desTime*1000)`	
`delay(delayTime);`	
`driveServos(0.0,0.0);`	

The last three lines deserve some comment. The lines before them set the motor drives and therefore their speeds. These last three lines specify how long the servos must spin. So we calculate the **delayTime** in milliseconds from our **desTime** (in seconds), and pass it to the **delay(...)** function which suspends the MPU for this time. Then, we drive the servos with driveL = 0.0 and driveR = 0.0 which will make them stop. So, the final result is that the robot will move 80mm forwards and it will take 2secs to do this, both values we specify.

Now the maths is perfect, and the code is perfect (they are both perfect *abstract* systems of thinking), but when you get the robot to execute this code, it will not move 80mm forwards, it will be more, or it will be less. Why? Because the robot is not *abstract*, it is not a *simulation*, it is really *embodied in the real world*. Think of its motors, their datasheet may specify their 'accuracy' as 10%. This means that if you ask them to rotate with an angular velocity of 10 rpm, they will rotate with anywhere between 9 and 11 rpm. In the worst-case scenario, the left motor could rotate at 9 rpm and the right at 11 rpm; the robot would not go straight forward but arc to the left! Also, we need to know the wheel radius, and so we measure it, but our measurements are subject to errors. And, also the relationship between **drive** and rpm may not be the one we presented earlier, each motor is different. Looking at the drive-rpm curve again (reproduced below) we see a real issue.

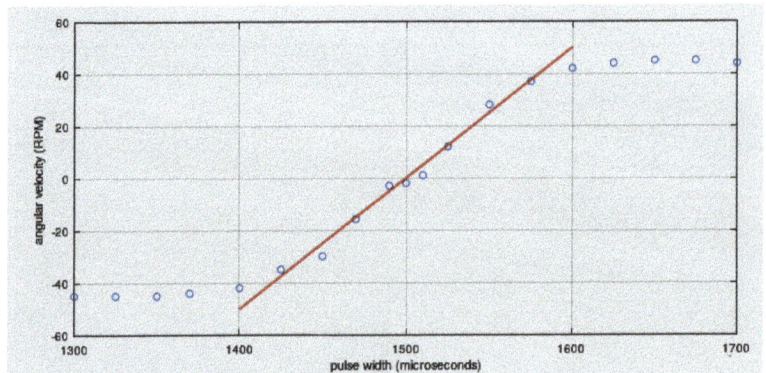

If we look at pulse Width 1500 (drive = 0) we see that increasing the drive to about 20 does not make the motor turn! There is a 'dead band'. This means that we cannot use small forward drives or motor speeds.

So, back to our problem. There are two ways to cope with this problem: The first relies on direct observation of the robot, and measurement of how it moves. Let's say we ask the robot to move 80mm and we measure how far it moves, 90mm; it's gone too far. So we could reduce the wheel velocity **omega**, but we know this is dangerous, since if omega is made too small, then the motor will not turn.

There is another way to make the robot travel less far; we can reduce the amount of time we drive the motors. So, we change the **desTime**

```
correctionFactor = desDist/actualDist;
desTime = desTime*correctionFactor;
```

It is important that we make this change at the appropriate place in the code, it must not be made before we calculate the omega's since they depend on the desTime. Here's where to put the correction, so that it only affects the **delayTime** which tells the motors how long to rotate.

```
correctionFactor = desDist/actualDist;

desTime = desTime*correctionFactor;
delayTime = (int)(desTime*1000)
delay(delayTime);
driveServos(0.0,0.0);
```

Figure 1.5 Wheel encoder: Top the IRed transmitter and receiver measures reflections from the spokes (bottom)

1.6 Wheel encoder technology

Here we shall look at improving our work on movement using dead reckoning by employing a robot *proprioceptive* sensor which measures the position of the wheels. The sensing device is called a 'wheel encoder', see Fig.1.5. As the wheel rotates, IRed light from the transmitter shown at the top passes through slots in the wheel, shown at the bottom. When the light hits a spoke, it is reflected to the IRed sensor and produces a pulse sent to the Arduino. So, as the wheel rotates, it sends a series of pulses to the Arduino shown in the diagram below.

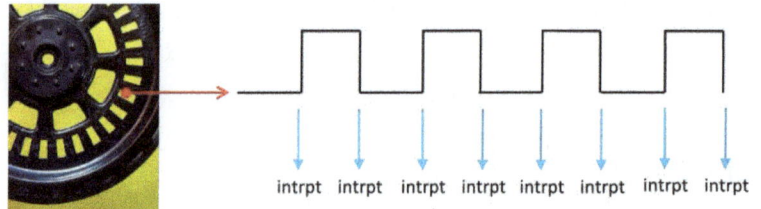

intrpt intrpt intrpt intrpt intrpt intrpt intrpt intrpt

The pulses arrive at Arduino pins especially configured to receive hardware 'interrupts' and when a pulse arrives, the code breaks out of its current execution place and jumps to an *Interrupt Service Routine (ISR)* and execution continues there. When the ISR is complete, then execution returns to the place where it was forced to break out from. Each wheel has an interrupt, the ISR for the right wheel is shown below.

```
void ISRoutineR() {
   countR++;
}
```

You can see this ISR increments the value of **countR** i.e., every time the encoder sends a pulse, this value is incremented, so the code knows how many steps the wheel has rotated. Here's an overview of how interrupts work. First the hardware pin is attached to the ISR, and the interrupt is configured to respond to a CHANGE (i.e., a *rising* or a *falling* pulse edge. This code is in **setup();**

```
attachInterrupt(digitalPinToInterrupt(EncoderR
_pin),ISRoutineR,CHANGE);
```

Here's what happens when the code is executing in loop() and a pulse arrives. Code execution is shown by the green arrow with a blob. When the interrupt is received, execution transfers execution returns to where it left off. It is important to understand that this occurs at the level of hardware instructions, not lines of code. It is extremely fast.

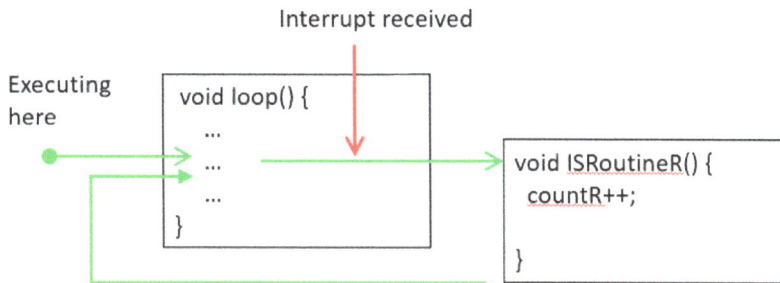

Now we need to understand how many pulses are received in one wheel revolution, and therefore how far the wheel moves between pulses. There are 32 spokes in the wheel, and since the ISR responds to both rising and falling edges, then there are 64 pulses per wheel revolution. The angle rotated between pulses is a little under 6 degrees, see Fig.1.6. But how far has the wheel moved? We know that

$$\Delta s = r\Delta\theta$$

so for a wheel radius of 33 mm, we have, calculating using radians

$$\Delta s = 33\frac{2\pi}{64}$$

Figure 1.6 Angle shown between pair of rising and falling pulse edges

which works out to be 3.24 mm. This is the accuracy of any measurement of wheel travel we can make since it is the smallest step in distance we can possibly know.

1.7 Moving on a straight line of given distance.

Let's say we want the robot to move a desired distance. How do we use wheel encoders to make this happen? First, we need to calculate the number of steps required, which is the number of pulses the encoders receive, setting **countL** and **countR**. The number of pulses is simply the distance gone divided by the step size,

$$n_L = n_R = \frac{desired\ dist}{\Delta x}$$

For example, to get the robot to move 300 mm, we need to count 300/3.24 = 93 pulses, approx. Here's some example code which will do the job. We set the motor drives and switch on the servos. Then we monitor the actual counts and when they exceed the counts we need ($n_L = n_R$) then we stop the servos.

```
driveL = drive;
driveR = drive;
driveServos(driveL,driveR);

if((countL > nL) && (countR > nR))    {
    driveServos(0.0,0.0);
    servosDetach();
}
```

1.8 Moving on an Arc

We've already seen the maths for this, but let's revisit it here. The arrangement is shown in Fig.1.7 where the symbol θ_{Rob} refers to the rotation of the entire robot about its centre (in the middle of its axle). The arc is specified by this angle and also the radius of the curve. Remember the robot axle has length $2a$ as shown in the diagram.

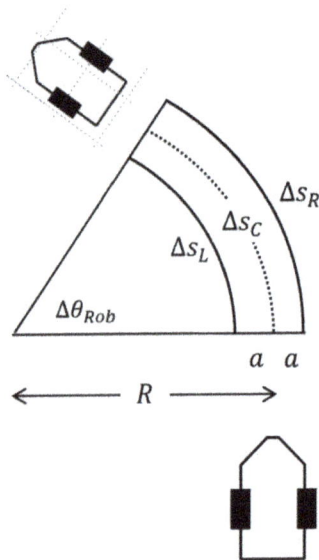

Figure 1.7 Geometry for moving on a curve with specified angle and radius.

The robot's wheels travel the following distances when traversing the arc,

$$\Delta s_L = (R - a)\Delta\theta_{Rob}$$

$$\Delta s_R = (R + a)\Delta\theta_{Rob}$$

The number of pulses is calculated by dividing the distances travelled by the step size Δx. So, we get

$$n_L = \frac{\Delta s_L}{\Delta x} = \frac{(R - a)\Delta\theta_{Rob}}{\Delta x}$$

and a similar expression for n_L. All quantities on the right are known, so it is easy to calculate the number of pulses for each wheel.

There is one additional factor we need to take into account. Not only is the right wheel making more steps, but it is also moving faster. Since both wheels must start and stop at the same time, the speeds are proportional to the number of steps. So, we have where *fwd* represents some *drive* to the servos,

$$\frac{fwd_R}{fwd_L} = \frac{n_R}{n_L}$$

The following code does all of this for us, where we are asking the robot to make a 90-degree arc of radius 300 mm.

```
desRadius = 300;
desDegrees = 90;
desTheta = desDegrees*(PI/180.0);

nLf =(desRadius - axleLen/2.0)*desTheta/dx;
nRf =(desRadius + axleLen/2.0)*desTheta/dx;

nL = (unsigned long) nLf;
nR = (unsigned long) nRf;
fwdL = 20;
fwdR = fwdL*nRf/nLf;
```

1.9 Robot Actuators

Robots move using their actuators (motors). We can classify actuators in several ways. First, we can consider how they move through space; most actuators we have experienced use rotation, which attached to wheels produce linear motion through space (along a straight or along a curve). But there are also *pure* linear actuators, which move in a straight line. Rotational actuators may be placed in sub-classes; there are *dc-motors* which when connected to a 5Volt source rotate with a certain angular speed, there are *servomotors* where you control their angular speed by giving them pulses of varying widths, and finally there are *stepper-motors* which move through a discrete angle when you give them a pulse. Stepper motors are capable of very accurate (can move small distances) and repeatable motion; they find application in laser-cutters, 3D printers, photocopiers and in robotic surgery.

Recent technology provides us with extended actuator possibilities, one example is shape-memory alloy. Think of a wire which can be long or short depending on its temperature; send a current through it (to heat it up) and it could be short, stop the current (it cools) and it will be long. So, you can control its length electronically. Applications are found in limb prosthetics.

1.10 Electro-mechanics of a Stepper Motor

The motor shaft is connected to an armature which rotates; think of this as having small magnets attached. Surrounding the armature are coils which produce a magnetic field when driven by a current. Fig.1.8 shows a simple example. At the top the armature is held in its position by the south pole coil attracting the north pole on the armature. Then the coil south pole is rotated clockwise, and a coil north pole is placed at the top, so the armature will rotate 90 degrees then stop there. So, steppers work by rotating a magnetic field around the armature, dragging it around. Clearly the rotation occurs in steps defined by the number of magnets.

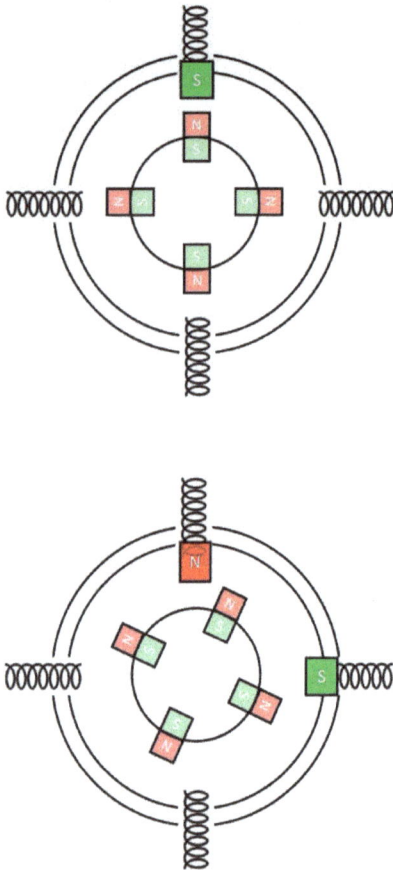

Figure 1.8 Stepper motor starting one clockwise step driven by magnetic attraction and repulsion

Our motors also have a gearbox which reduces the angle moved, this results in one revolution having 2038 steps. The angle of a single step (radians) is then

$$\Delta\theta_1 = \frac{2\pi}{2038}$$

which is about 0.0031 radians or 0.177 degrees, quite small. The distance covered by a wheel of radius r for one step is just

$$\Delta s_1 = r\Delta\theta_1$$

for our wheel of radius about 25mm this is around 0.08mm which is quite a small distance. This raises the hope of being able to control the position of a robot very accurately.

There are some limitations where this accuracy may not be realizes. One is related not to the motor, but to the wheels which may slip, especially if the motors are driven from rest to a high angular speed; to avoid this *speed ramping* will be used. Others are related to the motor, the motor may miss a step, especially when its speed is changed dramatically; ramping may help this. The other issue relates to driving two motors at the same time to make the robot move forward in a straight line. Assuming the following function is available (where the arguments are steps to the left and right motors respectively); compare the two coding solutions to get the robot to move *forwards* 100 steps.

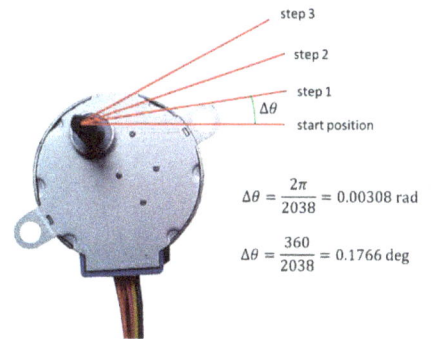

Figure 1.9 Our stepper showing a few anti-clockwise steps

`stepMotors(100,100);`	`int i=0;`
	`while(i<100) {`
	`    stepMotors(1,1);`
	`    i++;`
	`}`

The solution on the left will not work, since the internals of the function stepMotors(...) cannot drive both motors at the same time, and almost certainly drive one motor 100 steps then the other motor 100 steps. So, the robot would waddle forward in a series of arcs. While the same is true for the robot on the right, the waddle is limited to single steps and will be barely noticeable.

1.11 Forward Motion in a Straight Line

This is straightforward; we must arrange two things, (i) both left and right motors take the same number of steps, (ii) we must drive them with the same speeds. The API has two functions

```
setStepperSpeeds(speedL,speedR);

stepMotors(nL,nR);
```

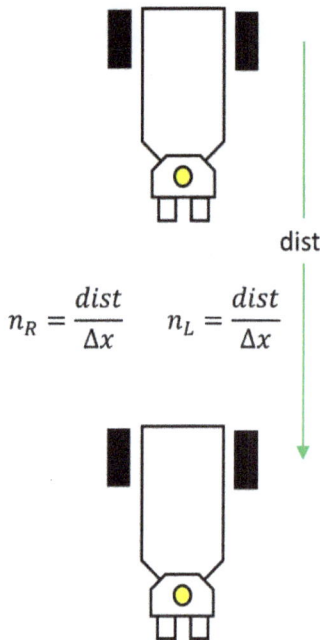

where **nL** and **nR** are the numbers of steps sent to the left and right motors. The calculation of these is straightforward; if we want to drive the robot **dist** forward then we have

$$n_L = \frac{dist}{\Delta x}, \qquad n_R = \frac{dist}{\Delta x},$$

where Δx (**dx** in code) is the distance travelled for one motor step. Fig 1.10. summarizes this.

1.12 Motion on an Arc

This is a little more complicated to analyse, but the implementation of the algorithm in code is really tricky and requires some thought. Let's say we want to move the robot along an arc of radius R and angle $\Delta\theta_{rob}$ in a clockwise manner. We have seen something similar before; the arrangement is shown in Fig.1.11. The left wheel moves further than the right wheel, and since both wheels must finish their journeys *at the same time*, the left wheel must move proportionally faster.

The distances travelled (Fig.1.11 top) are calculated as usual; for the left wheel we have

$$\Delta s_L = (R + a)\Delta\theta_{Rob} \qquad (1)$$

$$n_R = \frac{dist}{\Delta x} \qquad n_L = \frac{dist}{\Delta x}$$

Figure 1.10 Motion on a straight line; same speeds, same number of steps.

and for the right wheel

$$\Delta s_R = (R - a)\Delta\theta_{Rob} \quad (2)$$

Therefore, using the relation

$$n_L = \frac{\Delta s_L}{\Delta x} \quad (3)$$

and an equivalent one for the right wheel, we find

$$n_L = \frac{\Delta\theta_{Rob}}{\Delta x}(R + a) \quad (4)$$

and

$$n_L = \frac{\Delta\theta_{Rob}}{\Delta x}(R - a) \quad (5)$$

Everything on the right side of these equations is known, so we can compute the numbers of steps needed by left and right motors.

As mentioned, both motors need to complete their rotations at the same time, let's call this time Δt. Since distance is velocity time we have

$$\Delta s_L = v_L\Delta t, \qquad \Delta s_R = v_R\Delta t$$

and dividing these

$$\frac{\Delta s_R}{\Delta s_L} = \frac{v_R}{v_L} \quad (6)$$

and using expression (3) we have

$$\frac{v_R}{v_L} = \frac{n_R}{n_L} \quad (7)$$

so, the speeds are in proportion to the number of steps taken, in this case the right motor has a lower speed, agreeing with Fig.1.11.

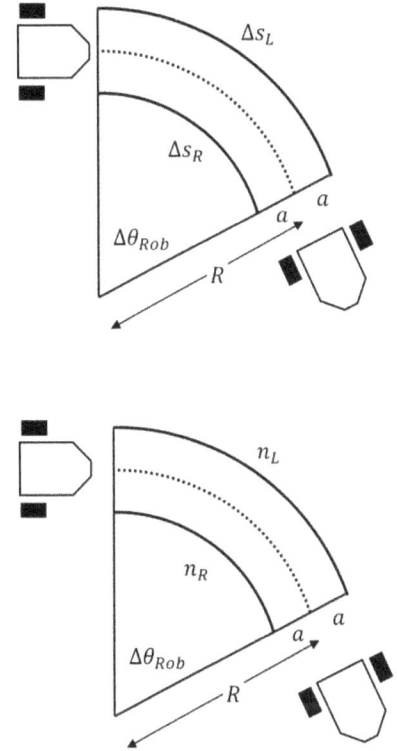

Figure 1.11 Driving on an arc: Top shows the distances, bottom expresses these in number of left and right steps.

We need to use expressions (4,5, and 7) in our code. These are used once to calculate the number of steps and the speeds of both motors. The problem is, once we have these values, how to use them to get the motors to turn correctly.

The approach we take (which is successful) is as follows. The right wheel moves a shorter total distance than the left wheel. So, when the right wheel has taken *one step,* we *imagine* that the right wheel has taken a *fractional step.* Of course, it can't but we imagine it can. When the left wheel takes another step, we imagine the right wheel as taken another fractional step.

At some time, the right wheel fractional steps will add up to a whole step, so at this point we make the right wheel take a step. How do we calculate the size of this imaginary step? Well, it is simply the ratio

$$\beta = \frac{n_R}{n_L} \quad (8)$$

so, when the left motor has finished its n_L steps, the total steps taken by the right motor is

$$\beta n_L = \frac{n_R}{n_L} n_L \quad (9)$$

which is just n_R, exactly what we want! Fig.1.12 shows this algorithm expressed as a flow diagram.

One solution to coding this is shown below. This assumes required step numbers and speeds have been computed. The loop runs over the left steps, and the variable **prepR** ('prepare the right motor') accumulates the fractional imaginary right motor steps, **beta** since $n_L = 1$ in (8). When **prepR** is greater than one step, the right motor is stepped. Note when the loop over the left steps is finished, there is a little check to see if there is an outstanding step for the right motor.

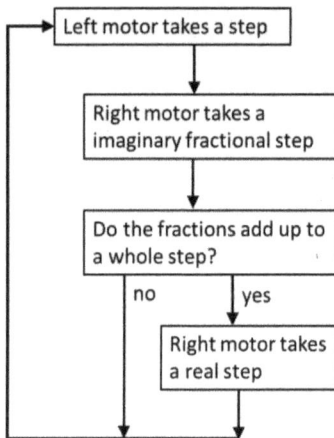

Figure 1.12 Arc algorithm

```
while (doneL < nL) {   // --------

   stepMotors(1,0);
   doneL += 1;
   prepR += beta;

   if(prepR > 1.0) {
       stepMotors(0,1);
       doneR += 1;
       prepR -= 1.0;
   }

} // End while ----------------

// Check if any outstanding Right
if (prepR > 0.5) {
   stepMotors(0,1);
   doneR += 1.0;
   prepR -= 1.0;
}
```

The variables **doneL** and **doneR** count the steps actually taken while **nL**and **nR** refer to the total steps to take. This code will only work for clockwise turns and needs to be generalized for all arcs.

Chapter 2
Robot Control Architectures

2.1 A brief Introduction

Robots are quite complex beasts; they are a mixture of mechanical systems, sensors, computational processes, and user interfaces. Moreover, robots live in a real physical (and not simulated) world, full of uncertainties and this world may change, e.g., as humans or other robots invade their space. So, you will not be surprised that the design of control systems or *architectures* can be quite involved. In this chapter we shall take a limited approach to explore architectures which you may experience on this module; we shall look at Finite State Machines, control algorithms for line following and wall-hugging (the so-called 'Bug' algorithm), and Rod Brookes 'subsumption' architecture. There are broadly speaking two approaches to robot control; the first is the **deliberative paradigm** and the second is the **reactive paradigm.**

The **deliberative** paradigm, which we won't be using, is shown in the diagram below. The robot must have an internal model of the world.

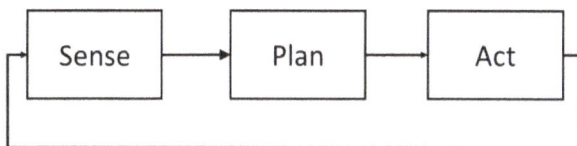

First the world is sensed, and then the internal model of the world is updated, and a plan is generated. This stage is computationally heavy since it involves automated reasoning based on the internal model and sensing. Finally, the robot acts. There are some problems with this model, first how to model everything the robot must know, while preventing the knowledge representation from becoming too complex. More importantly is the fact that sensing and acting are disconnected, the robot is unable to react to events such as imminent collisions. In the **reactive** paradigm, sensing and acting are closely coupled, so the robot can react to events. There is no planning and no world model, the robot uses the various behaviours, (move forward, turn, move back) and changes between behaviours based on the sensory input. This of course assumes that we can find meaningful 'primitive' behaviours on which the robot's actions can be based. Also, we need some sort of architecture to switch between these behaviours, and that's the focus of this chapter.

2.2 Finite State Machines

Here the behaviours of a robot are associated with 'states' in a finite state machine (FSM). An example of a FSM is shown in Fig.2.1 for a hypothetical drink vending machine; here drinks cost 30p and the machine accepts 10p and 20p coins.

The state names have been chosen to tell you how much money has been entered when the machine is in that state, so S20 means that in that state 20p has been entered. You start at the top in state S00 where nothing has been entered, then entering 10p will take you to state S10, or entering 20p will put you into state S20. The yellow highlights are *transitions* between states. There are three routes to get to the end state S30, entering 3 x 10p, or one 10p then one 20p or one 20p then one 10p. You can appreciate that this *state transition diagram* fully captures the state of the vending machine at each moment.

An alternative representation of a FSM is a table, which for Fig.1 looks like this. It shows the current state, the event triggering the transition and the new state

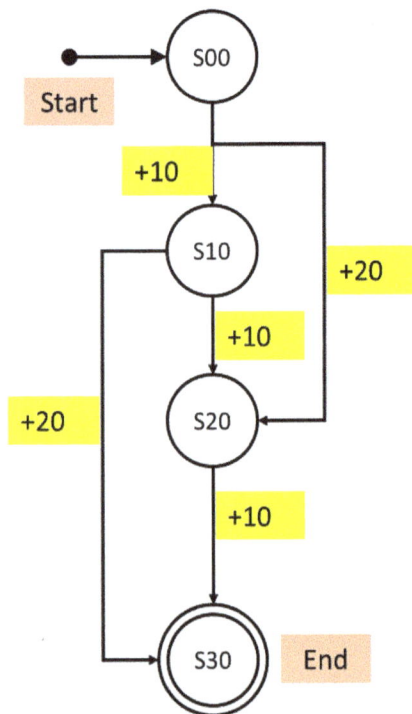

Figure 2.1 Finite State Machine for a vending machine

Current State	Event	Next State
S00	+10	S10
S00	+20	S20
S10	+10	S20
S10	+20	S30
S20	+10	S30

2.3 FSM for Robot Obstacle Avoidance

Have a look at the state diagram in Fig.2.2 Each state is labelled with a useful name. The line with the blob, top left, shows the starting state. Within the body of each state are the actions taken within that state. So for STATE_FWD where the robot is told to move forward, the values of driveL and driveR are set and then these values are sent to the servos. The bottom box of each state shows the *exit condition*, i.e., what causes the FSM to transit out of STATE_FWD. We look at the distance to the obstacle, when this is below a threshold we transit to STATE_BACK.

In the STATE_BACK, the drives are set and sent to the servo. After a time delay which we set then we make a transition to STATE_TURN. This state also drives the servos and exits after a timeout, and then transits to the first state STATE_FWD. Coding this FSM is straightforward. First, we define the states

```
#define STATE_FORWARD 1
#define STATE_BACKWARD 2
#define STATE_TURN_LEFT 3
```

These statements are used in the pre-compilation stage where any text `STATE_FORWARD` in your code is replaced by the value 1. Then the body of the FSM, which goes inside the Arduino's **loop(){ }** block looks like this, for the first state

```
switch(state)  {
    case STATE_FORWARD :
      driveL = 30;
      driveR = 30;
      driveServos(driveL,driveR);
      dist = getDistance();
      delay(60);
      if ( (dist < 300) && (dist != -1) ) {
        state = STATE_BACKWARD;
      }
      break;
```

You can see the lines to set driveL and driveR, to drive the servos, to get the distance, and the exit condition inside the

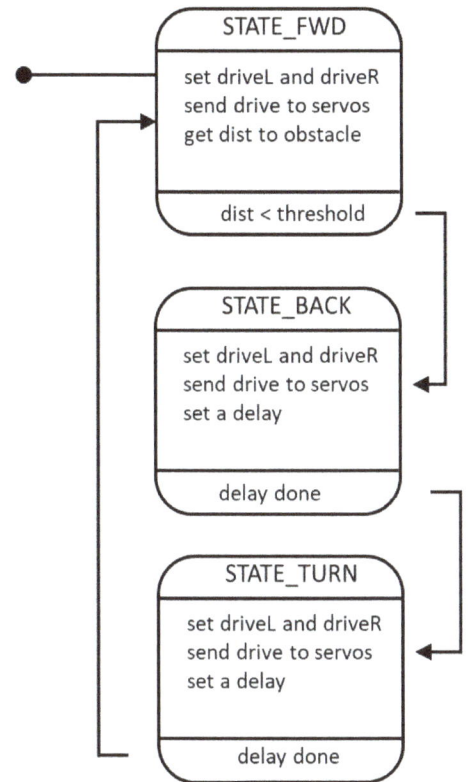

Figure 2.2 FSM for Obstacle Avoidance

if( ) statement. Note that a dist value of -1 means that the sensor has malfunctioned. The remaining states are

```
case STATE_BACKWARD :
    driveL = -30;
    driveR = -30;
    driveServos(driveL,driveR);
    delay(3000);
    state = STATE_TURN_LEFT;
    break;

case STATE_TURN_LEFT :
    driveL = 30;
    driveR = -30;
    driveServos(driveL,driveR);
    delay(3000);
    state = STATE_FORWARD;
    break;
```

This FSM is straightforward and does its job as you will see. Let's briefly look at another FSM, this time to keep a robot travelling along a line. The robot is our usual 2-wheeled differential drive critter, with two downward facing light sensors, see Fig.2.3; each sensor reports either 1 "I can see the line" or 0, "I can't see the line. So, we have a 2-bit binary system with 4 possible states:

Figure 2.3 Cute 2-wheeled differential-drive robot with 2 downward facing light sensors.

Sensors		State
Left	Right	
0	0	I'm confused
0	1	I need to turn clockwise
1	0	I need to turn anti-clockwise
1	1	I'm on the line

Have a look at the diagram below, where you can easily see all four states. From the above table and diagram, it's easy to code up a FSM which will do the job.

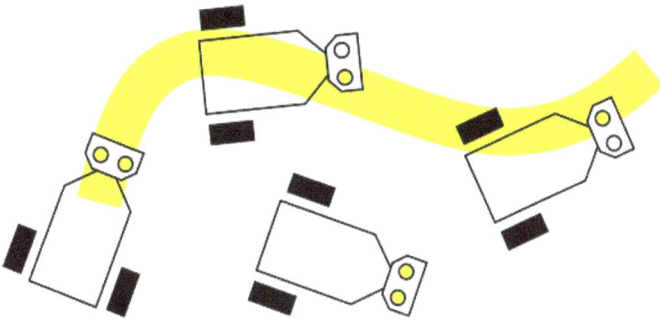

2.4 PID Controllers

Robots are often required to follow a line, e.g., in warehouse picking operations, or in auto-drive cars, or perhaps they need to navigate around obstacles using a 'wall-hugging' algorithm. We have just discussed how to create a controller using a FSM, and while this may work in many situations, it is actually a little crude. The reason is the motors will be driven with certain values of drive, but these are *fixed* such as {driveL = 0, driveR = 40} to get the robot to rotate anti-clockwise. We need a more delicate approach, where the further the robot if off the line, the more difference in drive we send to the motors. To do this we can use a PID ('Proportional, Integral, Derivative) controller.

Before we do that, let's look at a rudimentary (but acceptable) solution the problem; the configuration is shown in Fig.2.4, the robot has to get to the centre of the yellow line (on the red dashes) and the centre of its body is shown by the blue line fixed to its body. So here it is perfect. Now let's see when things go wrong. In the diagram below on the left, the robot is too far to the right. The error is the distance between the centre of the robot and the centre of the line (red dots). The sensor system will give us this error difference which we can use to correct the robot position by setting the motor drives. In the right part of the diagram the robot is still too far to the right, but not as much and the error is less. So, it makes sense to make the motor drives *proportional* to the errors, more error, more correction, and this is correct.

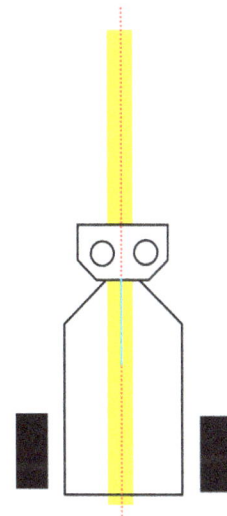

Figure 2.4 Configuration of a delicate line-follower

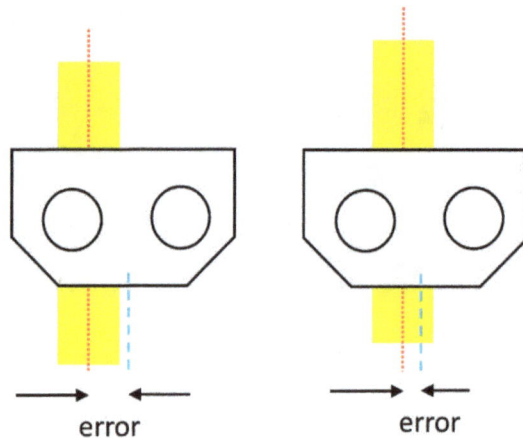

So how would we go about coding this behaviour? Here's how.

```
driveR = 20 + 10*error;
driveL = 20 - 10*error;
driveServos(driveL,driveR);
```

Here the '20' part of the drive, applied to both motors keeps the robot moving in a forward direction. We add an amount 10*error to the right motor to speed it up and subtract the same from the left motor to slow it down. So, the robot will turn anti-clockwise which is what we want. But the main point is, the larger the error, the more we add and subtract, to get the robot turning. This will work, but not always.

Where has the magic number '10' come from? To understand this, think what would happen if we replaced '10' by '1'? Well, the error would have only $1/10^{th}$ the effect, so the correction would not be as strong. So, the correction is *proportional* to this magic number 10. So let's replace this magic number by a coefficient Kp which stands for 'proportional coefficient'. We have just discovered the 'proportional' bit of the PID controller!

```
driveR = 20 + Kp*error;
driveL = 20 - Kp*error;
driveServos(driveL,driveR);
```

2.4.1 Why the Proportional Controller may fail

Here we shall start the development of the PID theory. Let's consider a toy problem where the robot has to move so that its lateral position is 1mm from the centre of the line. Let's try a few values of **Kp** to see how it fairs. Have a look at Fig.2.5 where the robot starts off at 0 mm and we drive it using the proportional error towards the 1 mm position (up the side of the graph). This is a graph of robot position against time, and we want the position to become 1.0, near the top of the graph. Horror! None of the values of **Kp** actually work! Also, for larger values of **Kp** the robot starts to oscillate! This is certainly not desirable. This is where PID control steps in.

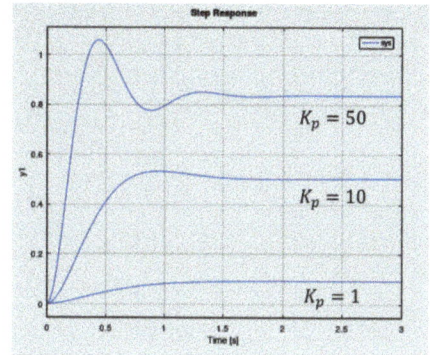

Figure 2.5 Response of robot to a disturbance with three values of Kp

2.4.2 Structure of the PID Controller.

This is shown in Fig.2.6 which shows the error signal $e(t)$ coming in at the left and the computed drive exiting at the right to drive out motors. The three boxes in the middle make different computations on the error signal, and these are summed (the Σ symbol) which form the final drive. We know how to calculate the proportional component. The integral component sums all the errors over time (this will include negative error values, so the sum will not simply increase). The derivative component takes the *difference* between current and previous errors. It turns out, applying some theory, that such an arrangement can provide good control of most systems, such as robots, self-driving cars, constant temperature heating systems, hard disk drive head positioning; you get the idea.

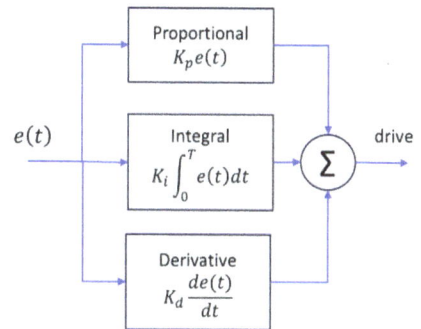

Figure 2.6 The structure of the PID controller; from top to bottom, P, I then D

So, let's see what the new components do, first the derivative. Let's stick with Kp = 50 (see Fig.2.5 where we had the horrible overshoot and oscillations) and look at two values of Kd; Kd = 0 and Kd = 10. You can see the results in Fig.2.7. With Kd = 0, we have oscillations and overshoot, but with Kd = 10, the oscillations are damped out, so the position rises smoothly towards the desired position 1.0, but still doesn't get there. That's the job of Ki as we shall now see.

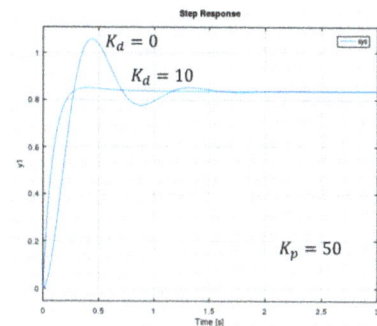

Figure 2.7 PID controller with two values of Kd

Have a glance at Fig.2.8 where the previous curve for Kp=50, Kd=10 (and assumed Kp=0) is drawn again, and the robot does not move to 1.0 mm but only manages a little over 0.8 mm. However, with Ki = 45 the results are much better, the robot is clearly moving towards the goal of 1.0mm.

So, in summary here is what the three coefficients do.

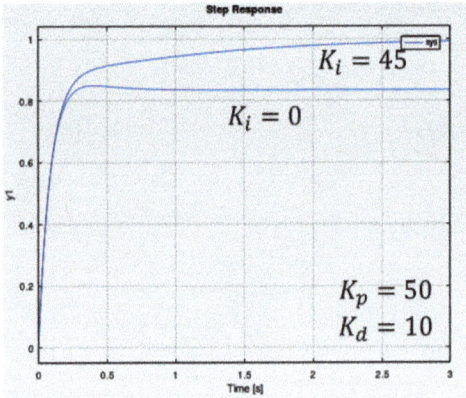

Figure 2.8 Effect of the coefficient Ki

K_p	proportional	Makes robot move to goal. Too large a value gives overshoot and oscillation. Goal is not achieved.
K_d	derivative	Overshoot and oscillation removed. Goal still not achieved
K_i	integral	Enables goal to be achieved.

2.4.3 Location of the Controller in the 'Loop'

Let's put all of this together and see where the robot fits in. Have a look at the diagram below which shows a general control loop which, as we have hinted above, can be applied to various devices, the 'plant', which is here our robot. At the left is the set point, in our example the desired robot position. At the right, a sensor monitors the actual robot position and sends this back to the start of the loop. Here the difference between desired and actual position is calculated to give the error signal which is then input into the controller.

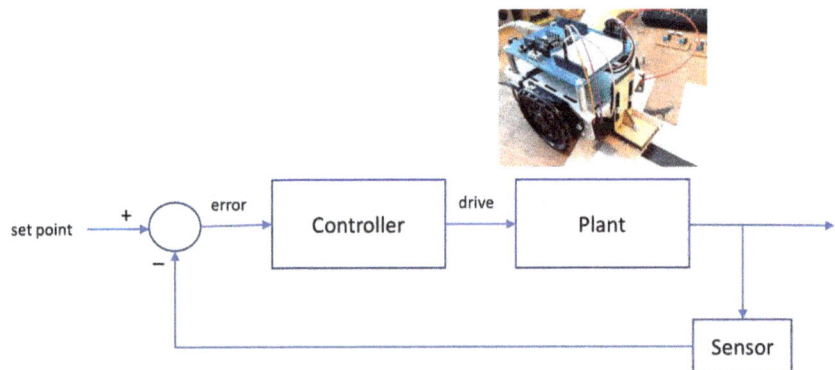

The PID controller does its job as described above and inputs a drive signal to the robot's motors making it to move towards the desired location.

While the *structure* of the PID controller will be the same for all robots, the coefficients will depend on the particular robot in question. Let's finally think about this.

2.4.4 'Tuning' the controller

This refers to finding the values of K_p, K_d, and K_i. This is usually done experimentally and is something of a black art which is learned by experience. The usual approach is as follows.

Stage 1.	Set all coefficients to 0
Stage 2.	Increase K_p until the robot shows signs of overshoot or oscillation
Stage 3.	Increase K_d until the oscillation disappears
Stage 4.	Increase K_i until the robot achieves the desired position. This may mean reducing K_p and perhaps K_i.

Chapter 3
Sensors

3.1 A brief Introduction

As humans our bodies are full of sensors, we depend heavily on sensing the external environment through vision, sound and touch. These sensors respond to the *external* environment and are responsible for giving us information about what is out there. This could be for navigation through a building, driving (while avoiding obstacles) reading a menu and making a choice of food to eat. Sensors which respond to the external environment are called **exteroceptive**. We also have built-in sensors which give us information about our bodies (and perhaps minds). We can sense when we have a toothache, when we are hungry, when we are tired or irritable. Sensors which respond to our internals are called **proprioceptive**.

The same classification is true for a robot, typical internal variables are battery voltage, motor speed, load on the wheels. External robot sensors can provide measurements of distance, sound amplitude and pitch and light intensity. Speed can be measured by doppler effect (change in pitch when a sound bounces off a moving object), and computer Vision can lead to object recognition (and therefore avoidance), path following and visual ranging. In this chapter we shall but take a small taste of what is possible and reflect on what we can achieve in our lab.

3.2 Active Ranging

3.2.1 Ultrasonic Pinging

The operation of the HC-SR04 ultrasonic ping detector is shown in Fig.3.1. This is somewhat like the principles used by bats in echo location. A distance measurement cycle begins by emitting a pulse of 40kHz waves from the transmitter Tx. This pulse spreads out (with a beam-width of

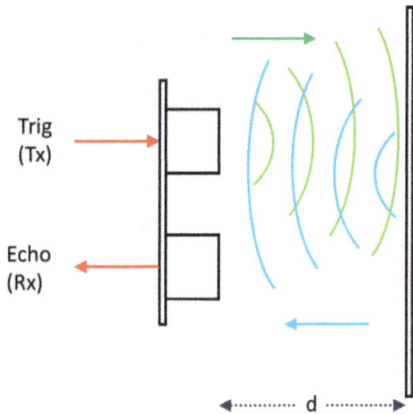

Figure 3.1 Ultrasonic Ping; green shows emitted pulse and blue reflected pulse.

about 30 degrees), and if it encounters an obstacle, then it is reflected and may arrive at the receiver Rx. The device gives us the time for the round trip which is (using the definition of velocity = distance over time)

$$t = \frac{2d}{v}$$

where v is the velocity of sound, around 330 m/s. Note the factor 2 since the round trip has distance *2d*. Fig.3.1 reminds us that we must *output* a trigger pulse to and *input* an echo pulse from the HC-SR04. Let's see the details and some code.

The diagram below shows the sequence of operations. At time 'A' we raise the value of trigger from LOW to HIGH and hold it there for 10μsec. Then, at time 'B' the device emits 8 ultrasonic pulses with frequency 40 kHz. When this is complete, the device raises its echo pin (which was LOW) to HIGH, telling our code the pulse has been emitted, time 'C'. When the pulse is reflected and enters the receiver, time 'D', then the ECHO pin goes low. So, by measuring the time the echo pin remains high, we know the time it takes for the pulse to do its round trip.

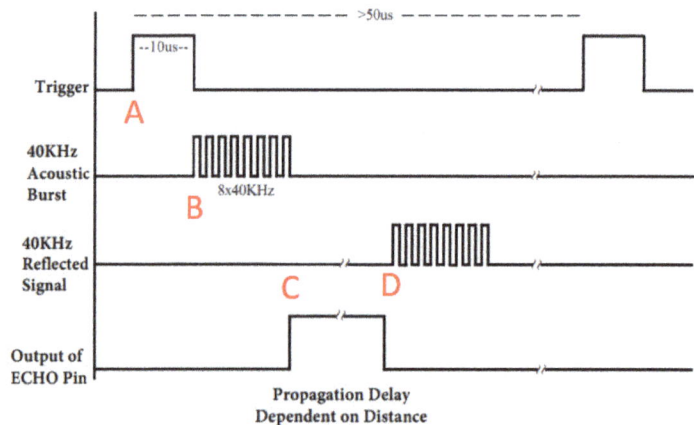

So, we can invert the above expression and calculate d.

$$d = \frac{vt}{2}$$

The code to do this is straightforward.

```
digitalWrite(HC_SR04_Tx,LOW);
delayMicroseconds(2);
digitalWrite(HC_SR04_Tx,HIGH);
delayMicroseconds(10);
digitalWrite(HC_SR04_Tx,LOW);
duration = pulseIn(HC_SR04_Rx,HIGH);
mm = 10*duration / 29 / 2;
```

The last line looks a little odd. This does several things; it converts duration (measured in microseconds) to seconds, then it converts the speed of sound 330 m/s to mm/s, and then it divides by 2 to take account of the round trip. The calculation has been done using integer arithmetic.

3.2.2 Limitations of Ultrasonic pinging

Think of an object around 2 metres away, this could easily be a wall in a room the robot must detect. It will take the ultrasonic pulse around 0.01 seconds (10 ms) to make its round trip. Our code must wait for this time to get the distance (unless we use interrupts), even then, the distance information is taking too long to be computed. A robot equipped with 10 such sensors arranged to measure distances all around its body must wait 0.1 seconds for this information to be available and moving at a leisurely speed of 100 mm/s, it would have travelled 10 mm and maybe suffered a collision. So, we must turn to faster approaches such as laser pings (LIDAR) which operate at the speed of light providing a speedup relative to sound of approximately $3 \times 10^8 / 3 \times 10^2$ which is about a million times.

3.3 Line Following

3.3.1 Aside – Resistors and Phototransistors

We have already encountered robot line following in Chapter 2 where we looked at the PID control algorithm. Now we need to focus on the details of the sensor system, so we can understand how to code the control algorithm. The following notes are from the point of view of an electronic engineer, so we have to get some basics in place, especially resistor circuits and circuits with a resistor and a phototransistor. Let's not worry about the theory, but rather look at some examples, plus a little dose of logic.

The starting point is *voltage*; we know that the Arduino can output a LOW signal (0 volts) and also a HIGH signal (5 volts) on a digital output pin. We also know it can input a signal on a digital input pin, and this can be HIGH (5 volts) or LOW (0 volts). So, voltage seems to be the key in understanding circuits. Now most input devices can be thought of as being configured as part of a *voltage divider*. This is shown in Fig.3.2 The rectangles are resistors, and these are connected between 5V and 0V (Gnd). There is an output voltage, so the question is what can this be? Well it can't be more than 5 or less than 0, since these values are not available at the input, so we conclude that the output voltage must be in the range 0.0 – 5.0. So, let's look at some concrete examples in the diagrams below.

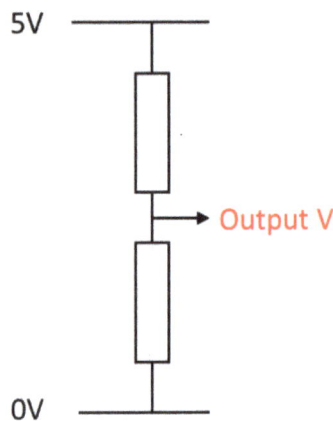

Figure 3.2 Voltage divider arrangement (resistor values not specified)

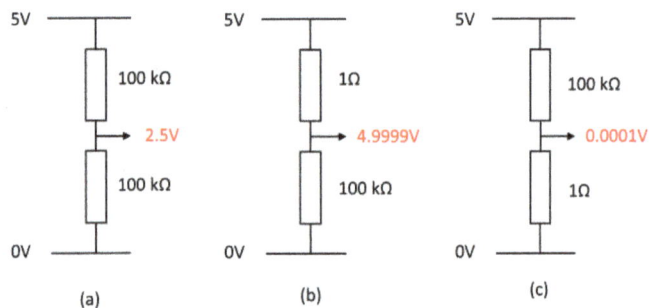

In (a) we two identical resistors, so *logic* tells us the 5 Volts must be divided into 2, so we get 2.5V out, it can't be anything else. In (b) we have a small resistor at the top and a huge on at the bottom, the voltage out is almost 5V. This makes sense since the output is 'more connected' to the 5V rail than to the 0V rail (less resistance). Conversely in (c) the output is 'more connected' to the 0V rail, so the output is close to 0V. Of course, you may have notice that the larger voltage is found across the larger resistor.

Now let's turn to the phototransistor which is used for most light-sensing activities, such as in our line following scenario, this is shown in Fig.3.3 where two situations are shown. On the left (a) there is a low light level applied to the phototransistor (two green arrows) and in (b) there is a higher light level (four green arrows). We need to *know* something about a phototransistor

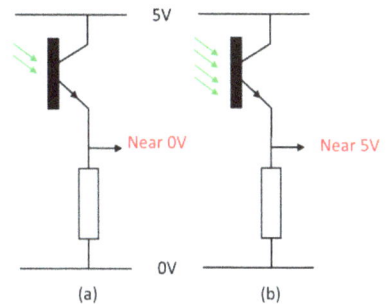

Figure 3.3 Phototransistor with (a) low level of light input, (b) high level of light input

---

A phototransistor is like a resistor whose resistance changes with its light input:
- Low light level – high resistance
- High light level – low resistance

So, in Fig.3.3(a) the phototransistor has a high resistance, so following our argument above, the output will be nearer 0V. In Fig.3.3 (b) with a high light level, the phototransistor has a lower resistance, so the output is 'more connected' to the 5V rail, so the output is close to 5V. In other words, the output of this circuit can give us a binary value 'there is light' (5V) or 'there is no light' (0V).

3.3.2 A Simple Line Detector
Here we shall briefly discuss a simplistic line detector, but it's not the one we advise to use, it's a toy problem just to reinforce some of the thinking we have presented above, as a stepping-stone to a more serious line detector. Let's assume we have a robot with a left and right sensor as described above, see Fig. 3.4(a). The robot is following a dark line on a lighter background. We want to know how the above

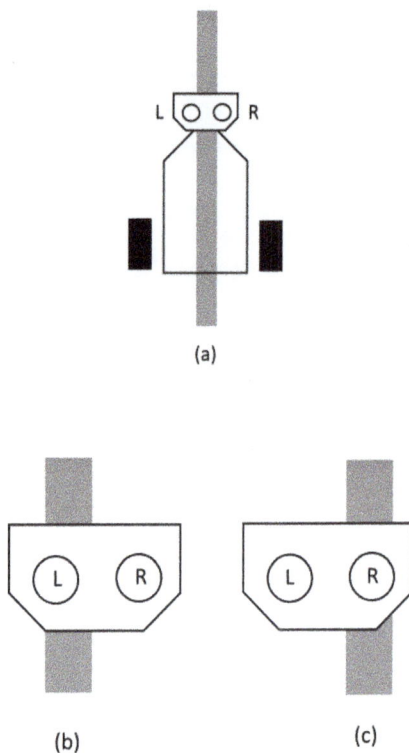

(a)

(b)   (c)

Figure 3.4 Simple line detection arrangement using phototransistors.

phototransistor circuit will give us information to keep the robot moving along the line.

In Fig.3.4(b) the robot has strayed to the right and must be instructed to rotate anti-clockwise. Let's have a look at the outputs of the L and R phototransistor circuit. The left one is darker, and therefore outputs a voltage close to 0V, and the right one is lighter, so outputs a voltage close to 5V. So, the left-right pair outputs {LOW, HIGH}. In Fig.3.4(c) we have the converse, the right phototransistor circuit is darker (outputting close to 0V) and the right one is lighter, (outputting close to 5V). So, the left-right pair outputs are {HIGH, LOW}. We can summarize this in the following table.

Situation		Sensors		Action
		L	R	
Fig.4(a)	on the line	HIGH	HIGH	continue forward
Fig.4(b)	off to the right	LOW	HIGH	rotate anti-clockwise
Fig.4(c)	off to the left	HIGH	LOW	rotate clockwise

This looks like potential input for a FSM as discussed in Chapter 2, providing the transit events between states. But as we also mentioned, this is a little brutal, and for smoother, more delicate control, we need a *continuous* input from our line sensor. So, let's see how we can achieve this. But first we need another step aside.

3.3.3 Aside – Capacitors and Phototransistors.

First let's turn to a capacitor; think of this like a bucket which accumulates charge when a current flows into it. When water flows into a bucket the water height increases and so does the pressure. When current flows into a capacitor the voltage (think pressure) similarly increases. Take a look at the diagram below.

In (a) the capacitor is empty, there is no charge so there is no voltage across it. In (b) current has started flowing into the capacitor, and you can see some charge has accumulated, so there is a small voltage across the capacitor.

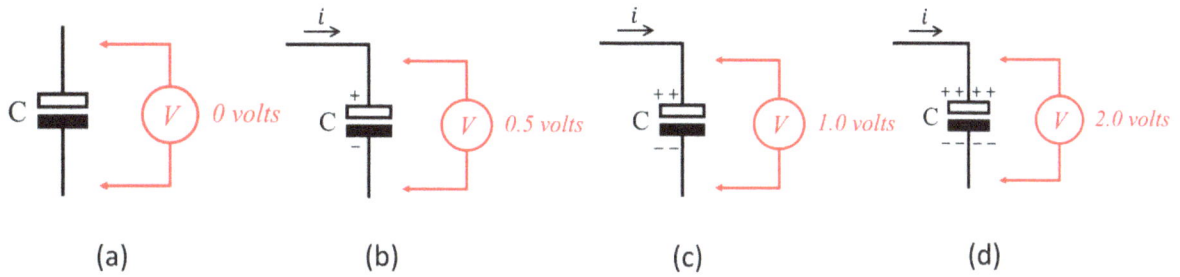

(a)   (b)   (c)   (d)

As the current continues to flow so does the charge accumulate (c and d) and the voltage rises. You can see that voltage is proportional to charge.

So, here's the circuit for our delicate line detector with an explanation of how it works. The blue arrow line is a connexion to an Arduino input/output. In (a) this is configured as an output and is given a HIGH level. Therefore, there is no voltage across and capacitor which therefore does not hold any charge.

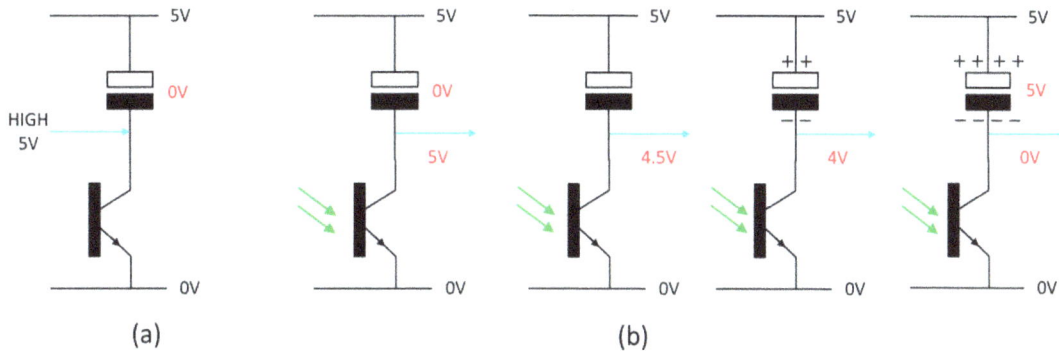

(a)   (b)

In (b) we see a time series of what happens when we remove the Arduino input, then the capacitor is free to charge, and as it charges, its voltage increases. So, the voltage across the phototransistor must decrease (since the voltage across both must always be 5V). Finally (rightmost circuit) there is 0V across the phototransistor.

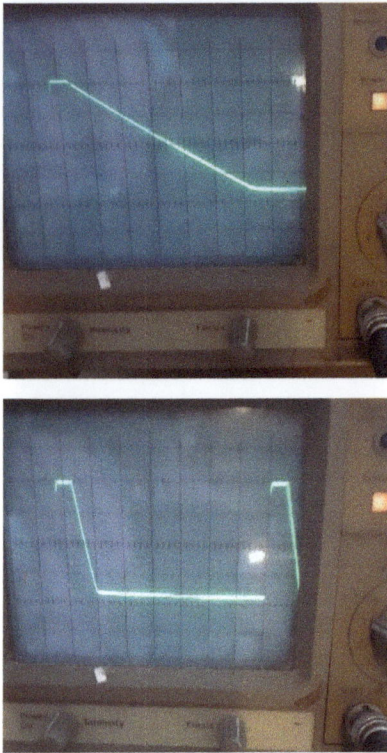

How does this help us create a delicate sensor to detect lines? Well, remember that the resistance of a phototransistor depends on the amount of light falling on it, more light means less resistance. So, in the above circuit, when the phototransistor gets a lot of light, its resistance is small, so it the current into the capacitor is larger, it charges up faster, and so its voltage rises faster. Therefore, the voltage across the resistor drops faster. This is the voltage we measure, and we measure the time it takes to drop to near 0V. A smaller time means more light, so the sensor is looking at white; a larger time means less light, so the sensor is looking at black.

Fig.3.5 shows some actual measurements I made on the circuit. You are looking at oscilloscope traces of the voltage across the phototransistor (vertical axis) against time (horizontal axis). The top photo shows what happens when you present a dark surface to the sensor and the bottom shows a light surface. The times to go from 5V to 0V are about

dark surface	3 μsec
light surface	0.5 μsec

Figure 3.5 Oscilloscope traces of sensor response: Top dark surface, bottom light surface

3.3.4 Coding all this Natural Nonsense

This all may seem complicated, perhaps you feel you have had a journey from Siberia to Nepal (via Hereford). So, let's look at some code, which may bring thinking together.

```
long RCTime(int sensorIn){
    long duration = 0;
    pinMode(sensorIn, OUTPUT);
    digitalWrite(sensorIn, HIGH);
    delay(1);
    pinMode(sensorIn, INPUT);
    digitalWrite(sensorIn, LOW);
    while(digitalRead(sensorIn)){
        duration++;
    }
    return duration;
}
```

We're looking at a function which returns the time **duration** for the circuit to respond to the light presented to the sensor. This code does the following

- Set the Arduino pin to output and set it HIGH. This discharges the capacitor.
- Wait 1ms to ensure the capacitor has had time to discharge.
- Set the Arduino pin to input and write a LOW to configure its internal operation.
- The while loop monitors the input voltage to the pin and loops until this is zero. During the loop, it increments the time **duration** variable.
- Finally, the function returns the value of **duration**.

In summary, our delicate sensor gives us a broad and continuous range of values from light (around 50) to dark (around 1000). This is an enormously useful range of values which we can easily use to compute the robot's error from the centre of a dark (or light) line.

Typical code to get the values from a left and right sensor and to compute some measure of error looks like this

```
senseL = RCTime(5);
senseR = RCTime(4);

error=((float)senseL-(float)senseR)/
((float)senseL + (float)senseR);
```

Note that senseL and senseR are **long** variable types and that these have been cast explicitly to **floats**. This is best programming practice. You may ask why the difference between the sensor reading is divided by their sum. Well, this is the expression

$$error = \frac{(sense_L - sense_L)}{(sense_L + sense_L)}$$

If the sensor values are the same (or close) this evaluates to close to zero. If one sensor value is large and the other is small then e.g., $(1000 - 10)/(1000 + 10) = 0.98$ which is less than 1. So this expression will produce error values in the range $0.0 - 1.0$. This value is 'normalized' and it is very, very, useful to know it is in this range.

Chapter 4
Navigation

4.1 A brief Introduction

We spend a lot of time walking around with a purpose; we have a goal, some place in space we must or desire to arrive at. That's in our mind, and everything else on our journey seems almost irrelevant; we negotiate paths, cross some roads, avoid walls and other folk unless we bump into a friend and stop for a chat; we may interrupt our individual goals, and decide to go for a coffee. To achieve this *navigation,* we use a combination of information; we may know the environment we are navigating through, the familiar corridors and rooms of a building we know. Also, we can adapt to changes in this familiar 'map' we have in our mind, when new furniture is placed along our way, and of course we can avoid other pedestrians.

Robot navigation is not too dissimilar; robots have a current position and a goal position, and our job is to devise algorithms to get them from the one to the other. Like us, they may have a *map* of their environment which they can use to navigate, and they should also be able to avoid obstacles. So, again there are two possible approaches to navigation we have seen before, a **deliberative** approach where the robot reasons using a stored map, and a **reactive** approach where it uses real-time sensor data to respond to obstacles and openings such as doors. These approaches may be called *planning* and *reacting*.

If the robot has a map of the environment, then the navigation problem is almost totally solved, however this may not be feasible due to constraints on robot memory and processing; this is especially true for the robots we are working with. Remember the microcontroller is dealing with many concurrent tasks; sensing the environment and driving its motors, so there may be little processing power left over to completely analyse a stored map and decide where to go. That's why we shall focus on *reactive* approaches.

To be able to navigate, our reactive robot needs to know exactly where obstacles are located in relation to the robot's *pose* (location and heading). The same is true for the detection of openings or gaps between obstacles. We shall start with these situations.

4.2 Object Localization

Consider the robot moving in the cluttered environment shown in Fig.4.1. We can see that the robot is moving towards a definite collision; it needs to *scan* the space it is moving into and detect the obstacle it is most likely to collide with first. So, we must equip our robot with some sort of scanning device, where it will scan the 180-degrees of space in front of it. We can choose various technologies for this scanner, laser, infra-red, but here we shall use our trusty ultrasonic device. The discussion is the same for other scanner types.

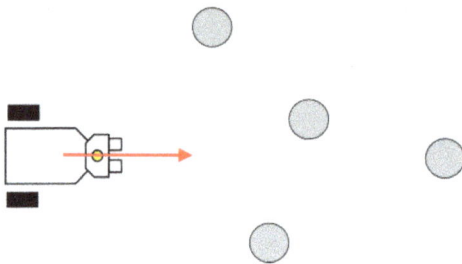

Figure 4.1 Robot moving to the right in a cluttered field of obstacles

Our robot is shown in Fig.4.2, the us-scanner is mounted on a small servo-motor which can move between 0 and 180 degrees, and we arrange that at 90 degrees it is facing forwards in the direction the robot is facing. Let's start with the simple problem of localizing a single object in the robots 'field of view'; so, the robot should find the angle of the object relative to its forward direction, and also the distance to the object. Then it can rotate so it is pointing at the object and then move towards it. This could form a useful application where the robot clears all objects from its arena.

The four stages of this algorithm are shown in Fig.4.3, first the robot (at rest) scans its environment and stores angles and distances to objects it encounters in an array. Then, still at rest it analyses the stored array and looks for the closest object and notes the angle of this object. Stage 3 it rotates so it is facing this object and finally in stage 4 it moves towards this object and pushes it out of the arena. Since there are four clear stages, we use a Finite State Machine architecture.

Figure 4.2 Robot equipped with an ultrasonic detector which can be scanned from right (0 degrees) to left (180 degrees).

The array shown below is a part of the entire array of 180 elements; the index into the array is just the current angle. So, in this example the smallest distance is 70mm at the array element 110. This is 110 degrees. In the third diagram we see the robot rotating from its current direction (90 degrees) to the target 110 degrees, in other words it must rotate anticlockwise through an angle 20 degrees.

The code to do this is straightforward; first we must declare an array of integers size 180, since we shall be scanning over a maximum of 180 degrees. The *index* into the array is just the angle.

```
float distArray[180];
```

Then we scan the turret and log the distances

```
for (angle = 20; angle <= 160; angle += 1) {
   servoTurret.write(angle);
   dist = getDistance();
   distArray[angle] = dist;
   delay(60); // Needed by US ping
}
```

Code to analyze the data, to find the closest distance *and to return the corresponding angle is*

```
for (angle = 20; angle <= 160; angle +=1) {
   if (distArray[angle] < minDist &&
distArray[angle] > 0.0) {
      minDist = distArray[angle];
      angleFound = angle;
   }
}
```

Figure 4.3 From the top: robot scans the environment and builds a distance array. It finds angle of closest distance and rotates to this angle. Then it moves forwards

Here is the array of distance values indexed by angle

160	170	170	200	300	350	360	350	360	350	320	100	90	80	70	80	100	410	410	450
96	97	98	99	100	101	102	103	104	105	106	107	108	109	110	111	112	113	114	115

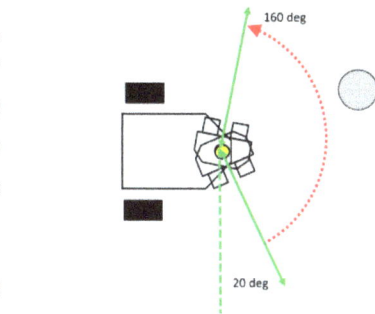

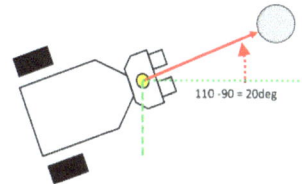

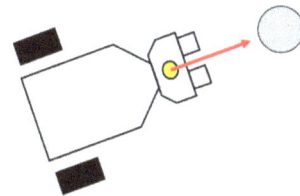

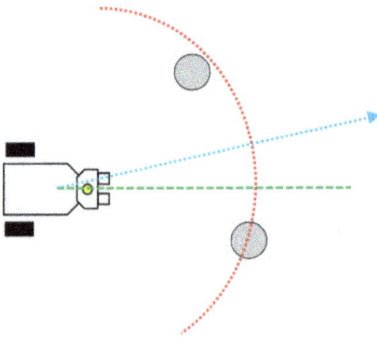

Figure 4.4 Robot detecting gap between objects.

4.3 Localization of Openings

Here we are looking at the situation where the robot must navigate to a target location and avoid all obstacles in its path; let's take the case of static objects. This is the situation shown in Fig.4.1. At any time, there will be two objects closest to the robot, so it makes sense to consider how to get the robot to pass through the gap between two objects. This is shown in Fig.4.4. The algorithm to do this is quite straightforward; as in the above example, the arena in front of the robot is scanned. But we do not need to save the distance information (at least present at each angle. We choose a threshold distance (red dotted arc in Fig.4.4) and log in a **labelArray[angle].** If the detected distance is less than the threshold, we put a **1** into the array, else we put a **0** into the array. Here's the code.

```
for (angle = 20; angle <= 160; angle += 1) {
  servoTurret.write(angle);
  dist = getDistance();
  if(dist < distThreshold)
    labelArray[angle] = 1;
  else
    labelArray[angle] = 0;
  delay(60);
}
```

A typical scan could produce an array like the one shown below. In this case the objects are located near 102 degrees and 112 degrees. The centre of the gap between the objects

0	0	0	0	1	1	1	1	0	0	0	0	0	0	1	1	1	1	0	0
96	97	98	99	100	101	102	103	104	105	106	107	108	109	110	111	112	113	114	115

is shown by the green arrow. The index (angle) of this is found by calculating the average of the angles when there is a label of 1. So, we calculate (100 + 101 + 102 + 103 + 110 + 111 + 112 + 113/8 = 107.

Having found this angle, we rotate the robot as in the above example and drive the robot through the gap.

4.4 Stopping at the Centre of the Gap

We might think of a situation where we want the robot to find the gap and move towards it but stop at the centre of the gap. This could be developed into an algorithm where the robot navigates through an entire *field* of objects, shown in Fig.4.5. The robot begins at the bottom by scanning and finding the first pair of objects; it moves to the centre of the gap and then stops. Then it performs another scan and finds the centre of the next pair of objects, moves there and repeats.

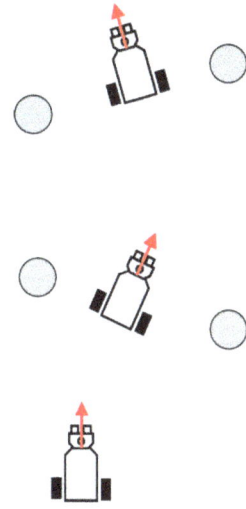

Figure 4.5 Robot navigating through a field of objects, in stages

The algorithm to do this is quite straightforward, in fact it is a simple extension (and enhancement) of the gap finding algorithm presented above. All we need to do is, in addition to logging an obstacle detected in the **labelArray[angle]** we need to record the distances in a **distArray[angle].** If we average the distances stored in this array, then this is a good approximation of the distance to the centre of the gap. So, the robot moves at the found angle, but in this case to the distance found. I mentioned that we need to *enhance* the gap-finding algorithm, and to do this we need to consider the geometry of the situation, shown in the sketch below.

The robot is located at the origin (0,0) and we extract from the distance array the distances to the inner edges of the objects, d_1 and d_2 and their corresponding angles θ_1 and θ_2. Our goal is to calculate the distance to the centre of the gap, d_C and the corresponding angle θ_C. We know how to get this angle, it is the average

$$\theta_C = \tfrac{1}{2}(\theta_1 + \theta_2)$$

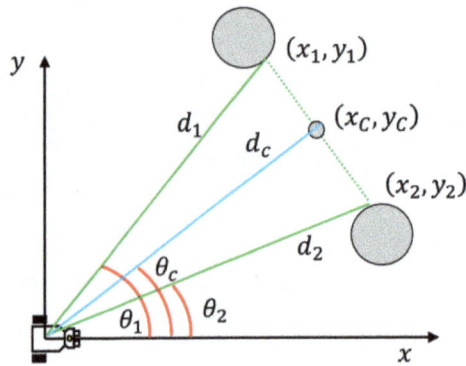

Now we must find expressions for (x_1, y_1) and (x_2, y_2). Using simple trigonometry, we have

$$x_1 = d_1 \cos \theta_1$$
$$y_1 = d_1 \sin \theta_1$$

and

$$x_2 = d_2 \cos \theta_2$$

$y_2 = d_2 \sin \theta_2$ Now, we know all distances and angles on the right of all expressions, so we can calculate (x_1, y_1) and (x_2, y_2). All we need to do now is to find the coordinates of the centre of the gap (x_c, y_c). Another diagram may help. Let's think about calculating x_c. We start off at x_1 and must

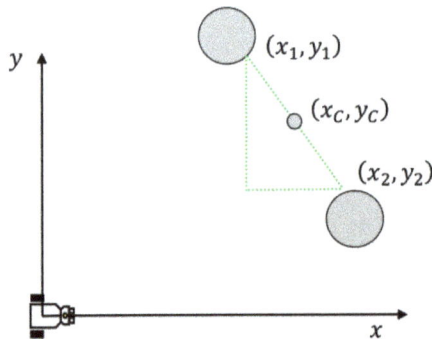

add half the width of the green dotted triangle

$$x_c = x_1 + \frac{(x_2 - x_1)}{2}$$

which simplifies to

$$x_c = \frac{(x_1 + x_2)}{2}$$

So, x_C is just the average of the bounding x-values. The same is true for y_C, so life has become quite easy here. Once we have (x_C, y_C) then we can calculate the desired distance

$$d_C = \sqrt{x_C^2 + y_C^2}$$

4.5 Other Approaches to Navigation

It's worth thinking about other approaches which are possible to implement within the constraints of our small robots.

The Bug Algorithm

This is the simplest object avoidance algorithm possible. The robot is instructed to move towards a target place, and if it encounters an obstacle on its way, it simply follows the contour of the object by 'hugging its wall' until it can see the target again.

Chapter 5
Robot Vision

5.1 A brief Introduction

Mobile robot perception is an interesting field of study and has evolved from ad-hoc solutions to specific robot situations to more grounded theory. Robots can be equipped with human-like senses (vision, sound, touch) but these can be supplemented with various others. Ultrasonic 'ping' rangefinders are perhaps inspired by the bat, motion detection by the fly eye. The compass could be likened to bird-brain sensory areas; it's interesting to look for natural analogues of other sensors such as GPS, wheel encoders, gyroscopes, laser rangefinders and doppler sensors.

Sensors can be classified as **exteroceptive**, those which respond to the external environment, such as vision, and **proprioceptive**, those which respond the robot's insides, such as battery voltage, wheel position and wheel load. Human vision is both a powerful sensory medium and is incredibly difficult to mimic in a robotics context; remember that over 50% of our brain is devoted to solving vision problems. Compared with other sensors, such as laser range-finding which responds to one (or a few) objects in a scene, robot vision has the potential to give information about the entire scene structure. The laser range-finder sends out a *ray,* and its collision with an object occurs at a particular angle and distance, whereas a camera has a *field of view* and can report all objects within that field. Usually images are processed before analysis; this may include edge-detection, segmentation and object labelling, or specific transforms which return information about straight lines, extracted by combing edges (Hough transform).

5.2 Limitations of Microcontroller Robots

First, we must accept the limitations of developing Computer Vision solutions for the small mobile robots, often based on Arduino technology we encounter. The first limitation is memory size. Consider a small image of resolution 300x200 with three colour channels, i.e., 3 bytes per pixel, which requires 180 kB of storage. The Arduino Mega2560 has 8 kB of data memory; clearly you cannot run image processing algorithms on this MPU, since there cannot be an image in memory! The second limitation is processing speed; take a 300x200 grey-scale image, performing a convolution with a 3x3 kernel, at a rate of 60 fps, requires a MPU clock speed of over 60 MHz whereas the Arduino gives us 16 MHz.

How can this be solved? Some companies offer Arduino-compatible alternatives with huge memory and fast processors (e.g, the Maixduino has 8 MB of data memory and runs at 400 MHz and retails at around £25). These boards mainly use the STMicroelectronics 'Cortex' MCU which is industry standard; the Maixduino board supplements this with a Kendryte AI processor. Compared with the Arduino, these boards are often tricky to bring into service, and documentation and blogs are hard to find, however we have had recent success getting the Maixduino up and running using PlatformIO. Then, of course, we could cross over to the dark-side and use a Raspberry-Pi, or even the NVIDIA Jetson technology.

Another solution is to off-load vision processing to a dedicated board, which applies one or more image processing algorithms, and sends the extracted features (such as segmented object sizes) to the Arduino for analysis. A feature can be coded in a few bytes, so memory space and transfer and processing rates is not an issue. This is the solution we shall encounter, our 'Pixy2' camera and processing board, which runs algorithms to (i) detect coloured blobs and return their location and size, (ii) detect lines, returning their endpoints as (x,y) coordinates in the image, (iii) detect types of intersections between lines. These are useful functions for a Robot Vision system, as we shall see. In addition, Pixy2 lets us extract individual pixels from the image, so we could just about code our own algorithm, e.g., a multi-line detector. This device is impressive, it boasts a dual-core 204 MHz NXP LPC4330 processor with an Aptina MT9M114 1296 x 976 resolution camera.

5.3 Pin-hole Camera

This is the simplest possible camera which you may have encountered in GCSE Physics and is a good approximation for many lens-based cameras. Look at Fig.5.1 showing a top view of a camera. Rays (green) from the red object pass through the camera iris (pin-hole) and form an image on the charge-coupled-device (CCD) retina. Sizes and distances are shown. The variable x is what we observe from the camera (and our code will report this). We need to know how to deduce the distance L of the object from the camera. We certainly do not know the value of d and we would like not to have to measure the width W directly.

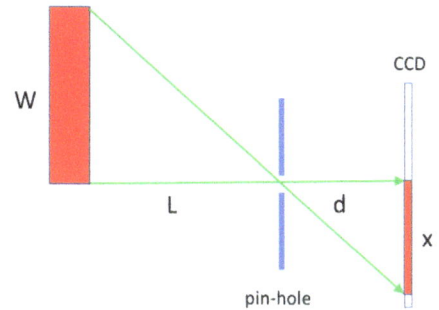

Figure 5.1 Pin-hole camera viewed from the top. Rays from the red object (width W) pass through the pinhole and create an image of size x on the camera's CCD retina.

But let's first remind ourselves of the geometry. Using similar triangles, we have

$$\frac{x}{d} = \frac{W}{L}$$

therefore

$$L = \left(\frac{Wd}{x}\right) \qquad (1)$$

This tells us that if we measure a small image width x then the object is far from the camera. Now, let's say we place an object at a known distance L_0 from the camera, and we measure the corresponding image size x_0, then substituting into (1) we have

$$L_0 = \left(\frac{Wd}{x_0}\right) \qquad (2)$$

and dividing (1) by (2) we find

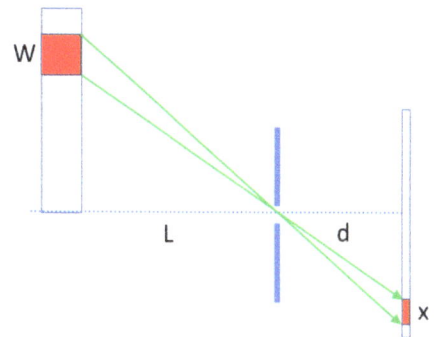

Figure 5.2 Arrangement we shall use in the lab, where the geometrical discussion is still valid.

$$L = \frac{1}{x}(L_0 x_0) \qquad (3)$$

This is useful, since the quantities in the bracket are known (we measure them), so we can deduce any distance L from the image width x, returned by our code. This is the process of *calibrating* our camera, preparing it for use. Note the units of the variables in (3). Both L and L_0 are measured in *physical* units (e.g., mm) but the x values are measured in pixels.

5.3.1 A Worked Example

Suppose we calibrate the camera. Assume the camera width resolution is 320 pixels. We choose to place the object so that its image completely fills the camera width. Let's assume we find this occurs at an object distance of 100 mm Then the above expression becomes

$$L = 32{,}000\frac{1}{x} \qquad (4)$$

Now we make a measurement of the image width x and we find this is 160 pixels. The distance to the object is (32,000/160) = 200 mm.

Now let's move the object and measure the image width x again, and say it has increased by the smallest amount, 1 pixel from 160 to 161. The object width is now (32,000/161) = 198.75 mm. This gives us the smallest measurable change in object distance for this situation, 1.25mm. Now let's investigate this, mathematically.

5.3.2 Sensitivity Analysis

It is useful to ask the question "how much does x change, when the distance to the object L changes?". This is one useful measure of the camera sensitivity. The quantity we wish to obtain is the relative (or fractional) change in x *to* L in other words

$$\frac{\Delta x}{\Delta L}$$

From expression (3) simple calculus tells us that

$$\frac{\Delta x}{\Delta L} = -\left(\frac{L_0 x_0}{L^2}\right) \qquad (5)$$

So, the sensitivity depends on L (in the denominator). For small values of L this sensitivity is large; a change in object distance will produce a larger Δx in the image width. This tells us that the camera is more sensitive to changes in object position L when the object is closer to the camera.

We can invert expression (4) and ask, "what is the smallest change in object distance which we can record in the camera image?".

$$\frac{\Delta L}{\Delta x} = -\frac{1}{x^2}(L_0 x_0) \qquad (6)$$

The smallest change in measured image width Δx=1 pixel. Using the values from our worked example above, $L_0 = 100\ mm$, x_0=320 pixels, and x=160 pixels we find

$$\Delta L = -\frac{1}{25600}(32000) = 1.25\ mm$$

This agrees with our worked example above. Perhaps this additional maths was not worth the effort.

5.4 Calibration of the Pixy2 Camera

The pinhole camera model presented above is useful in providing us with some understanding of the operation of a real camera. The actual operation of a real camera is best obtained using data from physical measurements. Here we report on calibration measurements for the Pixy2 camera, the experimental arrangement is shown in Fig.5.3 where the distance between the red object and the camera was changes (range 100 – 360 mm) and the image width in pixels measured. Since we know the relationship is inverse, see expression (3) then we plot distance versus 1/width. In other words, we are looking for the following *linear* relationship.

$$distance = m\left(\frac{1}{width}\right) + c \qquad (7)$$

Figure 5.3 Experimental arrangement to calibrate Pixy2

where m is the gradient of the straight line, and c is the intercept. Here's some typical results. The gradient is calculated as the length of the green arrow divided by the length of the red arrow (in units shown) and the intercept is the dist value where 1/width is zero on the plot

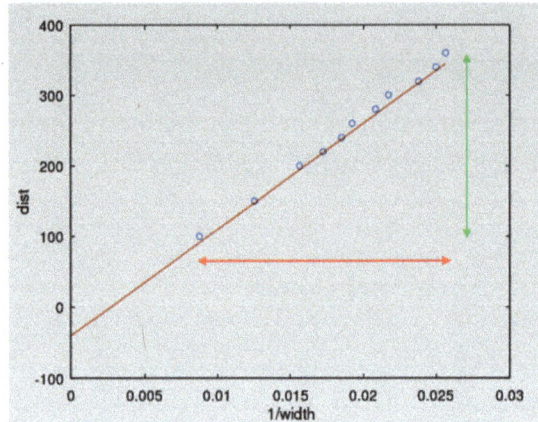

My estimates are: gradient = 15625, intercept = -40. So, the *approximate* relationship between width and distance is

$$distance = 15625 \left(\frac{1}{width}\right) - 40 \qquad (8)$$

However, we can do better than that. We can input the gradient and intercept *estimates* into a nonlinear regression program, which fits the curve to the data automatically, and gives us the optimal values for gradient and intercept.

5.4.1 Automatic Non-Linear fitting

This was done using the Octave script **PixyDist.m** which makes use of the function **nlinfit**. You need to provide a data set and a model to this function, here our model is the inverse relation between width and distance. The syntax for the model is

$$@(p,w) \ (p(1)./w) + p(2) \qquad (9)$$

The @(p,w) tells us that a function of variable **w** will follow where **p** are the parameters to be fit by the function. Running the script yields the following output

estimated parameters	15276.4
	-38.3
95% confidence intervals	14606.6 to 15946.1
	-51.5 to -25.1
r2 value	0.9966

The r2 value tells us that 99.7% of the data is explained by the fitted curve. The confidence intervals are fine, though the range for the second parameter is perhaps a little large. Our manual fit was not bad at all! The final relationship between width (pixels) and distance (mm) is therefore

$$dist = \frac{15276.7}{width} - 38.3 \qquad (10)$$

We can use expression (10) in our code. Just for completeness, here's the non-linear fit curve.

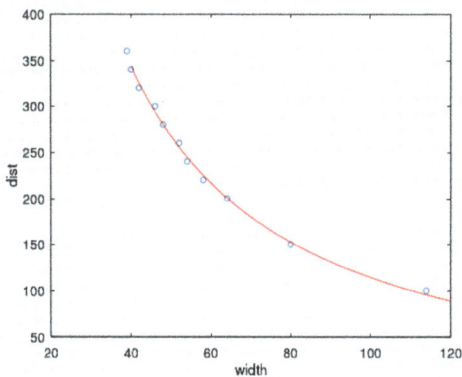

This non-linear curve fitting is a useful skill to have for other work. Now we can use the above values and write a function to convert image width to distance.

```
float getDistanceFromObject(uint16_t width) {
    float dist;
    dist = gradient/(float)width + intercept;
    return dist;
}
```

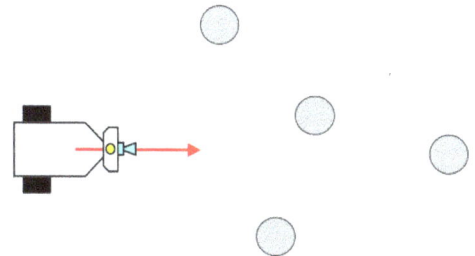

Figure 5.4 Robot moving through a cluttered environment, needs to localize each object so it can navigate between them.

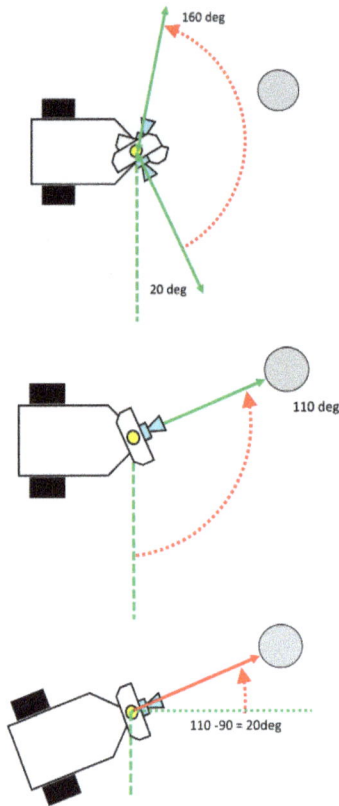

160 deg

20 deg

110 deg

110 -90 = 20deg

Figure 5.5 Robot scanning an environment. Top, scans, middle, finds an object at 110 degrees, bottom rotates to face the object ready for the kill

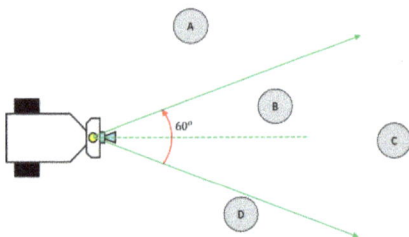

A
B
60°
C
D

Figure 5.6 Field of view of a camera, only objects B and C are perceived

We have managed to write a computational function which captures the workings of the camera based on experimental data.

5.5 Application – Object Localization

Object *localization* is more than object *detection*. In a detection situation, we are content with detecting that the robot is about to collide with something, so we can avoid it. Localization is more precise; when a robot *localizes* an object it finds out where it is (relative to its own location), in other words, it must find the angle of the object and the distance to the object.

When a robot moves in a cluttered environment (Fig.5.4) it needs to know where the objects are located. How it does this depends on its sensors. If it has a laser sensor, which sends out a **ray** which collides with an object, then it is clear that it needs to *scan* the space it is moving into. This means rotating the laser ray from 0 to 180 degrees (looking forward) and sensing any object at any angle. This is shown in Fig.5.5.

But when the robot has a camera, it may not need to do this scanning, since the camera captures objects within its *field of view*. The robot could simply analyse what it sees and based on this it would decide how to move.

Figure 5.6 shows such a scenario. Consider the case of a single object in the camera's field of view. The Pixycam can tell us the x-location of the object (measured horizontally from the left image boundary) and we can use this to generate an *error* signal to drive the robot wheels to move the object towards the centre of the FOV. This is shown in Fig.5.7 where the object is to the right of centre, so the robot has rotated *clockwise* in order to centre the object

If the camera is pointing forward, then the object is in the correct place when it is at the centre of the image; here there is zero error. We define a positive error when the object is too far to the right,

$$error = x - framewidth/2$$

First, we normalize the error, dividing the above by the value of $framewidth/2$. This results in an error which is in the known range of -1.0 to 1.0 irrespective of the image frame width. The drive signal to rotate the robot is taken (in the first instance) to be proportional to this normalized error. This whole algorithm can be seen in the following code snippet.

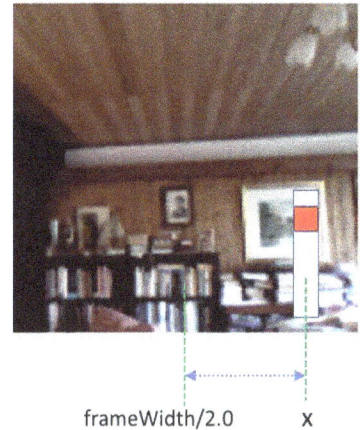

framewidth/2.0   X

Figure 5.7 Object to the right of centre. Its location x and frameWidth/2.0 define the error signal telling the robot how much to turn.

```
x = pixy.ccc.blocks[0].m_x;
error = ((float)x - (float)frameWidth/2.0);
error = error/((float)frameWidth/2.0);

driveR = -Kp*error;
driveL =  Kp*error;
driveServos(driveL,driveR);
```

The coefficient K_p is called the *proportional gain* for the above controller. We can estimate a suitable value. We know that the drives for our robot are around $20 - 40$, and we have normalized the error, so we find $K_p \approx 30$.

www.ingramcontent.com/pod-product-compliance
Lightning Source LLC
Chambersburg PA
CBHW042108210326
41519CB00064B/7588